Super-Duper Series

Marvelous Money!

by Annalisa McMorrow
illustrated by Marilynn G. Barr

Publisher: Roberta Suid
Design & Production: Scott McMorrow
Cover Design: David Hale
Cover Art: Mike Artell

Also in the Super-Duper series: *Incredible Insects!* (MM 2018),
Spectacular Space! (MM 2019), *Outstanding Oceans!* (MM 2020),
Wacky Weather! (MM 2057), *Peculiar Plants!* (MM 2058),
Amazing Animals! (MM 2059), *Leapin' Lizards!* (MM 2100),
Exciting Earth! (MM 2101), *Terrific Transportation!* (MM 2102)

Monday Morning Books
P.O. Box 1680
Palo Alto, CA 94302

E-mail us at: MMBooks@aol.com
Visit our Web site: www.mondaymorningbooks.com
Call us at: 1-800-255-6049

Monday Morning is a registered trademark of
Monday Morning Books, Inc.

ISBN 1-57612-113-5

Printed in the United States of America
987654321

Contents

Introduction: Why Money?

While money doesn't necessarily make the world go round, it is a concept that every child should master. With the activities in this book, your students will learn about the exciting world of finance, while practicing writing, reading, math, research, performance, and speaking skills. They'll interview a president, make wampum belts, write travel guides to famous monuments that appear on money, learn about the people on coins and bills, create copper plate money in the style of Pacific Northwest Native Americans, and much more. Most of the activities in this book can be simplified for younger students or extended for upper grades.

Marvelous Money! is divided into four parts and a resource section. Through a variety of activities, **Hands-On Discoveries** will help answer questions such as, "How are coins minted?" "What did people use before dollars and coins?" and "How does a checking account work?" The first part of this section features activities about objects that were used prior to modern money, such as wampum belts, manillas, and copper plates. This section also features math activities such as "Checking In." Reproducible sheets with a special dollar bill icon have directions written specifically for the children.

Nonfiction Book Links feature speaking, writing, and reporting activities based on nonfiction resources. Many activities are accompanied by helpful handouts, which will lead the children through the research procedure. When research is required, you have the option of letting children look for the facts needed in your classroom, school or local library, or on the Internet. Or use the "Super-Duper Fact Cards" located in the resource section at the back of this book. These cards list information for 16 money-related subjects. You can duplicate the cards onto neon-colored paper, laminate, and cut them out. Then store the cards in a box for children to choose from when doing their research. These cards also provide an opportunity for younger children to participate in research projects. Information is provided for them on easy-to-read cards.

The **Fiction Book Links** section uses chapter books and storybooks to introduce information about money. This section's activities, projects, and language extensions help children see connections between money and the world around them. Each "Link" also includes a tongue twister. You can challenge children to create their own twisters from the money facts and words they've learned. Also included in this section are decorating suggestions (called "setting the stage") for each particular book. Creating a book-friendly environment in the classroom will encourage children to read on their own for fun.

It's Show Time! presents new songs sung to old tunes. The songs can be duplicated and given to the children to learn. If you want to hold a performance, write each performer's name on the reproducible program page and distribute the copies to your audience. Consider having the children make costumes to go with the songs.

Each section in this book ends with a "Super-Duper Project," an activity that uses the information children have learned in the unit. These projects include "Put It in a Piggy" and "I'm on the Money." A choral performance is one possible "Super-Duper" ending for the "It's Show Time!" section.

Museums often feature exhibits of ancient money. At some point during this unit, consider taking a field trip to a museum or to a coin store. Or invite a coin collector to share his or her collection with the students.

All About Money

Coins or paper money haven't always been the methods for making payments. Early in history, people traded goods and services. Salt, tobacco leaves, feathers, shells, metal bracelets, and large stones with circles in the center have also been used at various times in history as ways to make payments or settle disagreements.

The first metal money was invented by Sumerians. They were the first to melt silver into small bars. Each bar was stamped with the exact weight to tell people how much silver they were getting or giving in return for goods or services. Gold has also been used for money throughout history.

When the Spanish conquistadors entered Mexico, the Mexicans were using cacao (chocolate) beans as payments.

Spanish silver coins could be cut into smaller pieces to make change. These were called "bits."

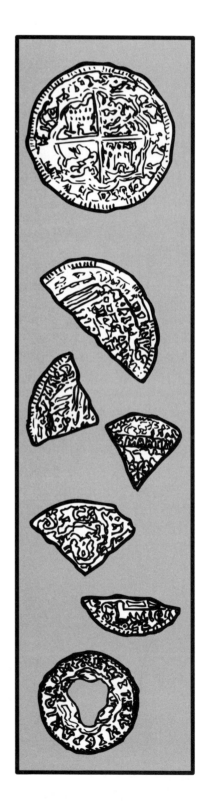

Marvelous Money © 2000 Monday Morning Books, Inc.

All About Money

Although paper money is common today, it differs from place to place. In the Canadian French colonies around 1685, money was printed in the form of playing cards. All four suits were represented. People must have had fun playing cards with their cash!

The United States now uses the same currency in all of its 50 states. This wasn't always the case. When the United States was a young country, different states had their own types of money.

In 1864, the secretary of the treasury ordered that the words "In God We Trust" appear on all coins minted during the Civil War. Today, this statement appears on all United States currency.

Wampum Belts

Native Americans from North America made belts of beads, called wampum, from white and purple clam shells. Indians used the clam shell beads as money, usually to settle disagreements between villages. Designs, such as squares within squares, were often worked into the belts. Shells were used as money in other places. In China, 3,500 years ago, cowrie shells were used. These shells were also used in India, Thailand, and Africa at different times.

Materials:
Purple and white construction paper, scissors, glue, tape

Directions:
1. Have the children imagine a time before people used metal and paper money. What might people have used in place of dollars and coins?
2. Discuss wampum belts with the children. Explain that the children will be making their own paper wampum belts.
3. Provide purple and white construction paper for the children to use to make representations of the wampum belts. They can cut designs from the white construction paper to glue onto the purple paper, or vice versa. They can tape sections of the construction paper together to make a piece long enough for a belt.
4. Post the completed belts on a "Wonderful Wampum" bulletin board.

Options:
• Before doing this activity, show the children pictures of wampum belts, such as the one in *Money* by Joe Cribb (Knopf, 1990).
• Bring in clam shells for the children to observe. Shells are often available from mail order catalogs.

Learning Chinese

In the early Chinese writing system, the cowrie shell was used as the symbol for money. The Chinese used these shells to make payments. Other cultures also used cowrie shells at different times. From the 10th to the 18th century, these shells were used in India. They were used in the 17th century in Thailand and in the 19th century in Africa.

Materials:
"Chinese Symbols" Hands-On Handout (p. 10), crayons or markers

Directions:
1. Duplicate a copy of the "Chinese Symbols" Hands-On Handout for each child.
2. Have the children study the symbols on the handout and then follow the directions on the sheet.
3. When the children are finished making their own symbols, let them share their symbols with their classmates. Ask the children to explain why they chose the symbols that they did.
4. Post the completed money symbols on a bulletin board in the classroom.

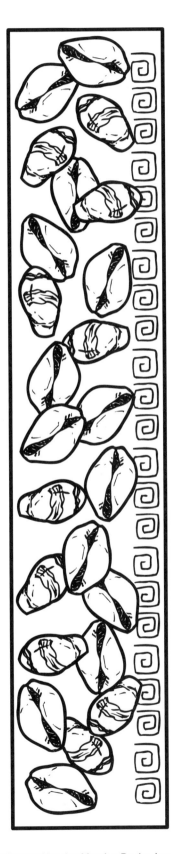

Marvelous Money © 2000 Monday Morning Books, Inc.

Chinese Symbols

Treasure	To sell	To buy	To barter

What You Do:

1. Look at the Chinese symbols in the boxes, above.

2. Use crayons or markers to make your own symbol for money.

_____

3. Make up additional symbols that mean the following:

Treasure:

_____

To sell:

_____

To buy:

_____

To barter:

_____

Marvelous Money © 2000 Monday Morning Books, Inc.

Making Manillas

In parts of Africa, before the introduction of paper and coin money, copper rings, known as manillas, were used to make payments. These large bracelet-like objects had a horseshoe shape. They came in assorted sizes and were made of a lead and copper mixture. Manillas were used in Nigeria until 1948.

Materials:
Aluminum foil or modeling clay

Directions:
1. Discuss the fact that today's money is usually carried in wallets, purses, or pockets. Then describe the manillas, which were bracelet-like items that could have been worn. How might that have made it easier for the users of manillas to make purchases?
2. Describe the manillas to the children. Then provide aluminum foil or modeling clay for the children to use to make their own manillas. They can make as many as they want in assorted sizes.
3. Let the children show off their manillas by wearing them on their arms.

Option:
• Before doing this activity, show the children pictures of manillas, such as the Nigerian copper manilla featured in *Money* by Joe Cribb (Knopf, 1990).

Creating Copper Plates

Prior to the arrival of European settlers, Native Americans used their own forms of money. Several coastal tribes in the Pacific Northwest used large hand-beaten copper plates for money. Animal faces were embossed into these shield-like plates, which could be as large as 3 feet high.

Materials:
Pacific Northwest Animals (p. 13), brown paper grocery sacks (one per child) or brown wrapping paper, crayons or markers, scissors, glue

Directions:
1. Cut apart the grocery sacks so that they lie flat, or cut individual pieces of the brown packing paper. The sheets should be approximately 2-feet by 2-feet, about as large as a cut-open grocery sack.
2. Duplicate a copy of the Pacific Northwest Animals for each child.
3. Discuss the Native American copper plate form of money.
4. Give each child a sheet of brown paper to crumple up and then smooth out. (This will give the paper a texture somewhat like hand-beaten copper.)
5. Have the children choose animals to feature on their plates. They can use the ones on the Pacific Northwest Animals patterns, or choose any others that they'd like. The children can cut out the patterns and glue them to their papers, or they can draw their own.

Pacific Northwest Animals

The animals that live in the Pacific Northwest include:

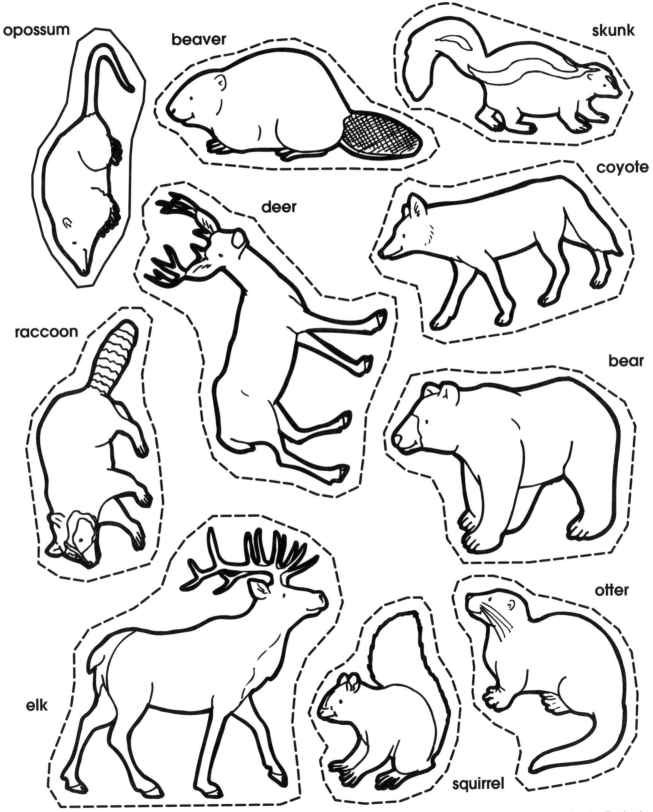

opossum

beaver

skunk

coyote

deer

raccoon

bear

elk

otter

squirrel

Worth Your Weight in Plates

The biggest coins ever minted were created in Sweden. The huge rectangular coins were called platmynt or plate money. The largest, the ten-Daler plate, weighed 42 lb. These coins were made between 1644 and 1809.

Materials:
Scale, calculators

Directions:
1. Describe the Swedish platmynt or plate money to the children.
2. Challenge the children to find out how many ten-Daler plates it would take to equal their own weight.
3. Provide a scale for the children to use to weigh themselves.
4. Help the children figure out approximately how much they weigh in ten-Daler plates. They can use calculators. For example, a child who weighed 80 lb. would weigh approximately two ten-Daler plates. A child who weighed 50 lb. would equal a little more than one ten-Daler plate. The children can round the numbers up or down to make them easier to work with.

Options:
• Younger children can simply decide whether they weigh more or less than one of the ten-Daler plates.
• The Native American shield-like copper plates were 3 feet high. Challenge the children to find out whether they are taller or shorter than this form of currency. Provide tape measures for the children to use to solve the challenge.

Minting Money

Modern coin money is made in places called mints. When a mint prepares to make a new coin, a design is chosen and a form is made. The metal used to make the coins is poured into the form when it's very hot.

Other types of materials are used in similar ways. For example, chocolate shapes are made by pouring liquid chocolate into a mold and letting the chocolate harden into the desired shape. Some artists make sculptures using similar methods. You can even make Jell-O shapes using molds! When the gelatin sets, it holds the mold's shape.

Materials:
Clay, wet sand, or gelatin; molds

Directions:
1. Choose a medium to work with.
2. Purchase inexpensive molds for the children to use.
3. Describe how coins are made to the students, and explain that they will be using the mold method to shape whichever material you've chosen.
4. Divide the class into small groups and let each group have a chance to make a mold of clay, wet sand, or gelatin.
5. Let all of the children observe each other's work. If you choose to make gelatin, serve it as a snack afterward!

Note:
If working with a gelatin recipe that calls for hot water, make the molds yourself at the front of the class and let the children observe.

Option:
• There are mints all over the world. Many of them offer tours to the public. Consider taking a field trip to a mint if there is one in your area.

Marvelous Money © 2000 Monday Morning Books, Inc.

Creating Currency

Students will be interested to know that people can't make their own money. This activity is called counterfeiting, and it is against the law. To purchase something in a store, you need to use actual money, not fake currency. However, students can make *pretend* money.

Materials:
Colored construction paper (red, green, yellow, purple), scissors, crayons or markers, envelopes (one per child)

Directions:
1. Have the students imagine that they have each started their own country. What type of money would they need? Would they want paper money as well as coins?
2. Ask the students to think of the different shapes that money comes in: rectangular paper and circular coins. Do these shapes make sense? Why aren't there triangular bills? The students can use a variety of shapes to design their own currency.
3. Have the children all use the same colors of paper. They should make at least the following amounts: one red bill, two green bills, three yellow bills, and four purple bills. All of the children should have an assortment of colored bills when they're finished. (These bills will be used in a later activity.)
4. Children can store their completed currency in the envelopes.

Note:
Save this money for the "Weighing Money" (p. 17) activity.

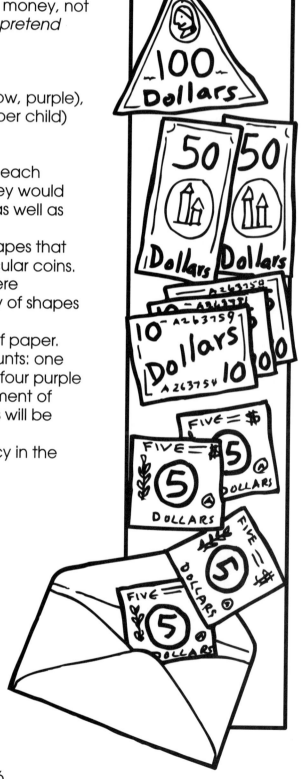

Weighing Money

A United States dollar bill doesn't weigh very much by itself. It takes nearly 500 dollar bills (490 bills) to make a pound. In this activity, the children will work together to find out how many of their own bills equals a pound.

Materials:
Balancing scale, money from "Creating Currency"

Directions:
1. Explain how many United States dollar bills it takes to equal a pound.
2. Bring out the balancing scale and have the children each add their own currency from "Creating Currency" to the scale until it reaches a pound. If there isn't enough paper money to make a pound, the children can see how many of their bills it takes to make a half of a pound or a quarter of a pound. Then work together as a class to figure out how many bills it would take to equal a pound. (If they have enough money to make half of a pound, they would double that number. If they have enough to make a quarter of a pound, they would multiply that number by four.)

Options:
• Bring in rolls of pennies and have the children see how many pennies it takes to make a pound, half pound, or quarter of a pound.
• It takes 233 bills stacked together to equal one inch. Have the children see how much of their own currency it takes to equal one inch.

What's It Called?

Different countries have different names for their monetary units. For example, Australia, Canada, Hong Kong, New Zealand, Taiwan, and the United States all use the word "dollar" to describe their money. Austria uses the word "schilling," Belgium and France use "franc," the Czech Republic uses a "koruna," and Malaysia uses a "ringgit."

Materials:
Money from "Creating Currency" (p. 16), map or globe

Directions:
1. Explain to the children that different countries have different names for their money.
2. Share the list of names (below) with the children. Point out the different locations on a map or globe as you tell the children the name for each country's currency.
3. Have the children take their own currency out of the envelopes and look at it.
4. The children can name their currency using a word that is already used by a country or by making up their own word for their newly minted money. They can choose a name for their currency based on the size, shape, or color, or they can simply name it for their own unique reasons.
5. Have the children present their money, and their money's new name, to the rest of the class.

Note:
For additional names, look in the foreign money section of your local newspaper.

Foreign Money Names

Brazil: New Cruzado
China: Yuan
Ecuador: Sucre
Finland: Markka
Greece: Drachma

Japan: Yen
Jordan: Dinar
Spain: Peseta
United Kingdom: Pound
Venezuela: Bolivar

The Money Exchange

Money from different countries equals different amounts. For example, at certain times one U.S. Dollar may equal six French franc. On the same day, the U.S. Dollar might also equal 1800 Italian lira. These rates can change daily. To find out the exchange rate, check the business section of the newspaper, look online, or contact a bank.

Materials:
Money from "Creating Currency" project

Directions:
1. Explain the concept of an exchange rate to the students.
2. Write the following Exchange Rate on the chalkboard for the children: One red bill equals two green bills. One green bill equals three yellow bills. One yellow bill equals four purple bills. From this rate, the children can also figure out that one red bill would equal six yellow bills, and so on. (Simplify the exchange rate for younger students by making one color of a bill equal to either one or two of another color.)
3. Have the children decide whether they think the exchange rate is fair. If not, have them work together to create their own exchange system. They might work with the colors of the bills, or with the sizes and shapes of the money.
4. Divide the children into small groups. Have the groups use the money from the "Creating Currency" project to trade with each other.
5. Ask questions about the procedure. Was it difficult to get everyone to agree upon the rate? Have the children imagine how people from different countries might feel when they exchange money.

Option: Cents and Sense
• Challenge the students to solve this riddle. A famous folk tale tells the story of two workers who helped a king. The king asked how the workers would like to be paid. The first worker said he would like 10,000 dollars. The second worker said she would like a penny for one day, two pennies the next day, four pennies the third day, with the pennies continuing to double every day for 30 days. Which worker made the better deal? Why?

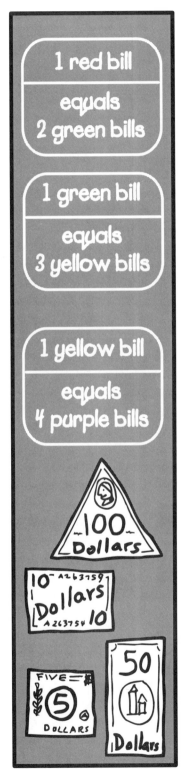

Checking In

Checkbooks are a way of spending money when you don't have actual currency with you. A check is an I.O.U. that says that you promise to pay the amount of money written on the check. The first checks were simply pieces of paper that stated money was owed.

Materials:
"Checkbook and Register" Hands-on Handout (p. 21), construction paper, stapler, scissors, pens or pencils

Directions:
1. Bring in a real checkbook and register for the children to observe.
2. Duplicate a "Checkbook and Register" Hands-on Handout for each child.
3. Have the children cut out the checks and put them in construction paper checkbook holders (construction paper cut into rectangles and folded in half). The children can decorate their holders.
4. On the insides of the checkbook holders, the children should glue the registers.
5. Tell the children the amount of their starting balances. The children should write this number on the registers inside their checkbooks.
6. Draw items on the chalkboard and label the items with dollar amounts. Have the children decide which items they would like to buy. Have them fill out checks for these items and note the check number and amount in their register.
7. When the children are out of money or checks, they should turn in their checkbook to be corrected.
8. Continue this activity through the year, providing additional checks and registers for the children to use.

Notes:
• For younger children, the dollar amounts for the items to be purchased can always be even. Older children can work with decimal numbers. For instance, an item can cost $2.75 or .62 cents.
• Advanced students can write checks for large amounts and use calculators to check their balances.

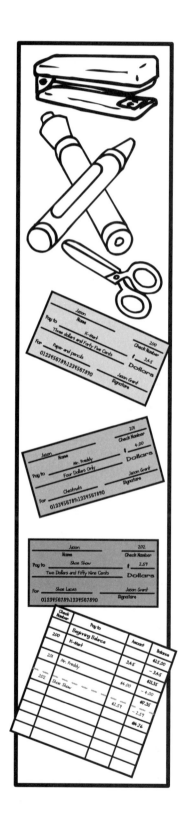

Checkbook and Register

Check Number										
Balance										
Amount										
Pay to										
Check Number										

Name _____ Check Number _____

Pay to _____

$ _____

_____ Dollars

For _____ Signature _____

0123456789:1234567890

Name _____ Check Number _____

Pay to _____

$ _____

_____ Dollars

For _____ Signature _____

0123456789:1234567890

Name _____ Check Number _____

Pay to _____

$ _____

_____ Dollars

For _____ Signature _____

0123456789:1234567890

Name _____ Check Number _____

Pay to _____

$ _____

_____ Dollars

For _____ Signature _____

0123456789:1234567890

Beyond Money

What did people do before they had money? Often, they traded. People might trade goods (actual items) or services (things they could do). In fact, when Columbus discovered America, he was looking for an easier trade route to India. He thought he found India when he landed in America. That's why he named the people he met Indians!

Materials:
"Cool Coupons" Hands-On Handout (p. 23), construction paper, scissors, crayons or markers

Directions:
1. Duplicate a copy of the "Cool Coupons" Hands-On Handout for each child.
2. Discuss the concept of trading with the children.
3. Have the children fill out the answers to the questions on the handout. Then provide art materials for the children to use to make Cool Coupons.
4. Once all of the children have completed the assignment, set up a trading time when the children can check out each other's coupons and set up trades. If children don't want to trade, they can give their coupons to their families or friends for presents.

Cool Coupons

Before people had money, they traded goods (actual items) for services (things they could do).

What You Do:

1. Imagine you are a farmer and you need to go to a doctor. How could you pay the doctor?

_____

_____

_____

_____

2. Think of things you could trade with your friends. For example, if you are good at making birthday cards and have a friend who knows how to play football, you might trade cards that you make for football lessons. Write down skills that you have that you feel are worth trading.

_____

_____

_____

_____

3. Make coupons that are good for the skills that you have. Trade your coupons with your friends for skills that they have.

Rupee Rubbings

Coins from other countries are fascinating to both children and adults. When friends travel, ask them to bring back coin money from other countries for you to borrow. Or get foreign coins from a bank. You might also ask coin collectors to let you borrow interesting-looking coins that are not too valuable. Try to find as many different types of coins as possible. Otherwise, use coins from your own country.

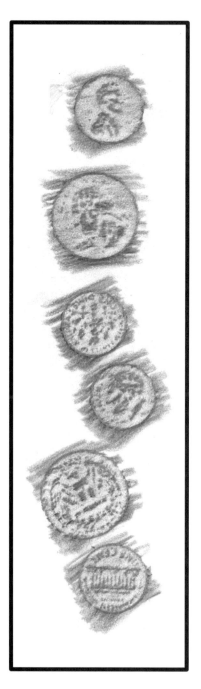

Materials:
"Rupee Rubbing Wallets" Hands-on Handout (p. 25), thin white paper, crayons, coins (from abroad, or from your own country; enough for each child to use several in his or her art), stapler

Directions:
1. Duplicate a copy of the "Rupee Rubbing Wallets" Hands-on Handout for each child.
2. Provide the art materials for the children to use to do this art activity. (Directions for the activity are on the following page.)
3. When the children are finished making their wallets, they can share them with the rest of the class.

Option:
• Plastic coins with raised patterns are available from many teacher supply stores. They come in assorted denominations and can be used in place of real coins.

Note: Rupees are the form of currency in India.

Rupee Rubbing Wallets

What You Do:

1. Spread out the coins beneath a piece of paper.

2. Do a rubbing over the coins using the flat side of a crayon. Use different colors to make a more interesting picture.

3. Fold the paper into thirds, as shown. Staple two sides shut. You can use this as a wallet to store your money (real or play).

Activity:

1. Draw a picture, or do a rubbing, of your favorite coin.

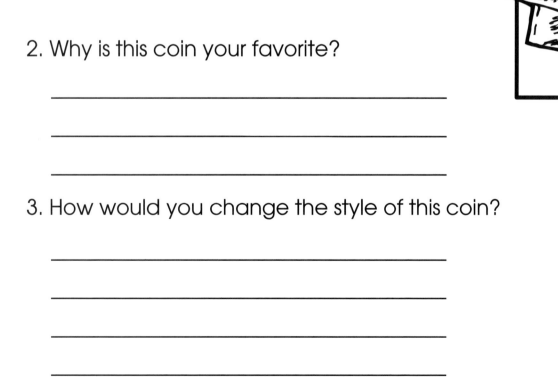

2. Why is this coin your favorite?

_____

_____

_____

3. How would you change the style of this coin?

_____

_____

_____

_____

A Quarter 'Til

This is a fun time-telling activity with a verbal tie-in to money. To emphasize the focus on fractions, remind the children that four quarters equal a dollar, *and* four quarters of an hour equal an hour. This activity may be an extra challenge to children who are used to telling time via digital clocks or watches. Make sure to create a clock for yourself.

Materials:
Clock (p. 27), scissors, brads, crayons or markers, glue

Directions:
1. Duplicate a copy of the Clock for each child and punch a hole in the center of each Clock.
2. Have the children cut out the clock hands and attach the hands to the clock face using a brad.
3. Let the children decorate their clock faces using crayons or markers.
4. Set the clock hands so that they are at a quarter before the hour. Have all of the children set their clocks, as well. Ask the children what time it is. Explain that often when it is 15 minutes before an hour, people say that it is "A quarter 'til" the next hour. When it is 15 minutes past an hour, people say that it is "A quarter past."
5. The children can continue to practice their time-telling skills, focusing on a quarter 'til and a quarter past the different hours.

Clock

Put It in a Piggy

Once the children have learned about different types of money, both foreign and domestic, they may be more inclined to save their own money.

Materials:
Flour, water, shallow pans for papier-mâché mix, balloons, newspaper ripped into long strips, tempera paint, paintbrushes, pipe cleaners, knife (for adult use only)

Directions:
1. Ask the children if anyone has ever told them to save their money for a rainy day. What do they think this means? Why would someone want money when it rains? Most people think that this statement means they should save money for a time when they don't have a lot of it.
2. Make a papier-mâché paste by mixing flour and water together. The paste should be slightly watery.
3. Have the children make their own piggy banks by covering a blown-up balloon with papier-mâché. Demonstrate how to dip strips of newspaper in the papier-mâché and wrap them around the balloon. The children can fashion small legs and a snout.
4. Let the banks dry.
5. Provide tempera paint for the children to use to paint their banks pink. They can glue on cardboard ears and add a pink pipe cleaner tail.
6. Cut a slit in the top of each bank. Then give the banks to the children to use to save their money for a rainy day!

Option: Make a Money Monster
• Banks can come in lots of different shapes. They don't need to be in a piggy bank shape. Using a wide-mouthed jar with a lid, the children can make money monsters. They can screw the lid on or off when they want to take money out or put money in. The jars can be covered with papier-mâché and painted. These monster banks also make great presents!

Interview with a President

American currency usually features pictures of presidents. The only coin that doesn't show a president is the Susan B. Anthony dollar. Several bills feature men who weren't presidents. Alexander Hamilton, first secretary of the treasury, is on the ten-dollar bill. Benjamin Franklin is on the 100-dollar bill. Salmon P. Chase, the 25th secretary of the treasury, is on the 10,000-dollar bill.

Materials:
Money Patterns (p. 30), "President Fact Sheet" Hands-on Handout (p. 31), "President Interview Sheet" Hands-on Handout (p. 32), pencils or markers, Super-Duper Fact Cards (pp. 70-77)

Directions:
1. Duplicate one copy of the Money Patterns for each child.
2. Have the children choose a president to research.
3. Duplicate one copy of the "President Fact Sheet" handout and the "Interview Sheet" for each child.
4. Have the children research the presidents using the guidelines on the "President Interview Sheet" handout. They can use the "Super-Duper Fact Cards" at the end of the book, or they can use books from the library.
5. Once the children finish their research, divide them into pairs. Let each partner take a turn interviewing the other in front of the class.
6. Set up an interview schedule, perhaps working through five or six interviews per day.

Option:
• Children can dress up to look like their chosen presidents. For example, a child pretending to be Lincoln might wear a tall paper hat and a fake beard.

Note: Children can choose to research Susan B. Anthony, Alexander Hamilton, Salmon P. Chase, or Ben Franklin for this report.

Money Patterns

One-dollar bill

George Washington

Five-dollar bill

Abraham Lincoln

Ten-dollar bill

Alexander Hamilton

Twenty-dollar bill

Andrew Jackson

Fifty-dollar bill

Ulysses S. Grant

One-hundred-dollar bill

Benjamin Franklin

President Fact Sheet

Use this fact sheet to record at least four facts about your chosen president. (Remember to list the books you used. If you use fact cards, write "fact card" under "Books I used.") Use the back of this sheet if you need more room.

My name is: _____

My president is: _____

Fact: _____

_____

Fact: _____

_____

Fact: _____

_____

Fact: _____

_____

Books I used:
Title: _____

Author: _____

Title: _____

Author: _____

George Washington was the first president of the United States.

President Interview Sheet

Write your answers under the questions. Write your own question for question 5. Or use a new sheet of paper to write and answer all of your own questions. Your partner will use these questions to interview you in front of the class.

Question 1: What is your name?

_____

Question 2: Which type (or types) of money features your picture?

_____

_____

Question 3: What is one important thing you did as president?

_____

_____

Question 4: What is on the back of the money that features your picture?

_____

_____

Question 5:

_____

_____

_____

The Flip Side

The pictures on the back of coins and dollars are sometimes as interesting as what is on the front. In this activity, the children will learn about the buildings featured on the back of different coins and dollars. Then they will write travel guides for future visitors to these sites.

Materials:
Behind the Money (p. 34), "Money Travel Guides" Hands-on Handout (p. 35)

Directions:
1. Duplicate a copy of the Behind the Money patterns and the "Money Travel Guides" Hands-on Handout for each child.
2. Let each child choose a place to research from the Behind the Money patterns. They can choose from American money or use books to research places on foreign bills or coins.
3. Have the children research their chosen place using the guidelines on the "Money Travel Guides" handout. They can do their research using the "Super-Duper Fact Cards," books, encyclopedias, or the Web.
4. When they are finished with their research, they can write brief travel guides as described on the "Money Travel Guides" handout.
5. Post the completed travel guides in the classroom where all of the children can observe them.

Option:
• Discuss the symbolic meaning behind the designs on the back of the one-dollar bill. The bill features the Great Seal of the United States. One half shows the "Eagle of Democracy." The other shows a pyramid. The pyramid represents strength and duration.

Behind the Money

One-dollar bill

**Great Seal of the U.S.
with pyramid**

Five-dollar bill

Lincoln Memorial

Ten-dollar bill

U.S. Treasury Building

Twenty-dollar bill

The White House

Fifty-dollar bill

U.S. Capitol Building

One-hundred-dollar bill

Independence Hall

Marvelous Money © 2000 Monday Morning Books, Inc.

Money Travel Guides

Use this sheet to record at least three facts about your chosen location. Then write a fact of your own.

The name of my location is: _____

1. In which state is the building located?

_____

2. Why was this building built?

_____

_____

_____

3. When was the building built?

_____

_____

_____

4. _____

_____

_____

Once you have written the facts, create a travel guide by drawing a picture of your chosen building on a sheet of paper. Write the facts about the building beneath the picture. Add any information you think a visitor would find interesting.

Marvelous Money © 2000 Monday Morning Books, Inc.

Learning the Lingo

Dollars, cents, and cash are all words to describe money. There are many other words, as well. Here are American slang terms that refer to money: moolah, silver, bread, cabbage, lettuce, dough, cash, dinero, and green. Other countries have different terms. For instance, in Germany, some slang terms for money are pebbles, charcoal, and moss.

Materials:
Money Patterns (p. 30), "Weird Word Hunt" Hands-on Handout (p. 37), pens or pencils, paper, student thesauruses

Directions:
1. Duplicate a copy of the "Weird Word Hunt" Hands-on Handout and the Money Patterns for each child.
2. Ask the children if they know any other words that mean money.
3. Have the children look at the Money Patterns and try to come up with their own descriptive terms for the different bills. They should also come up with new names for coins.
4. The children should write a reason next to each name they come up with. For example, someone might call quarters "roughies," because their edges are serrated.

Option:
• A large vocabulary will make your students' writing more interesting. Have the children write a short story in which the characters use slang. (It is more conventional to use the words in dialogue than it is to use them in narrative. For example, a character might say, "I'm out of dough." However, it would sound strange to read: He had no cabbage in his wallet. Or: The ATM machine was out of bread.) Explain the rules at the start of the assignment—no obscenities!

Weird Word Hunt

What You Need:
Pencil or pen, thesaurus

What You Do:
1. You are about to embark on a weird word hunt! Look at the words below and find as many different ways as you can to say the same things.

Child _____

Happy _____

Laugh _____

Pretty _____

Wet _____

If you get stuck, look up the words in a thesaurus. (This is a book that has many words with the same meaning.)

2. In a notebook, make your own thesaurus by collecting words that have the same meaning. You can start by keeping a list below.

_____

_____

_____

Marvelous Money © 2000 Monday Morning Books, Inc.

Match the Money

Materials:
"Match the Money" Hands-on Handout (p. 39), crayons or markers, "Where in the World?" Hands-on Handout (p. 40)

Directions:
1. Duplicate one copy of the "Match the Money" Hands-on Handout for each child.
2. Have the children try this matching game either by themselves or in small groups. (Directions for the activity are on the following page.)
3. Go over the answers to the matching game as a class. (The answers are upside-down on the bottom of this page.)
4. When the children are finished matching the money, give them each the "Where in the World?" Hands-on Handout. Explain that they will each be choosing one of the countries from the "Match the Money" handout and doing research to learn at least two facts about that country. For instance, a child who chooses Australia might learn about one animal and one type of tree or flower found in Australia.
5. Post a world map in the classroom and let the children point to the country they've researched as they share the facts they've found.

Koalas can be found in Australia.

Answers: 1H, 2E, 3D, 4G, 5B, 6C, 7A, 8F

Match the Money

Different countries use different forms of money. If you visited another country, you would need to trade your money for the other country's money. You could do this at a bank. When you returned home, you would change the foreign money back to your own money.

What You Do:

1. Pretend that you are traveling to the countries below. Try to match the money to the countries.

1. Australia	
2. United Kingdom	
3. Portugal	
4. Luxembourg	
5. Belgium	
6. Greece	
7. Canada	
8. Japan	

A

B

C

D

E

F

G

H

2. Choose one of the countries above to research. Write the name of the country below. You will be doing research to find two facts about this country. You might learn what type of food grows in the country or what types of animals live there.

Where in the World?

Name:

_____

Date:

_____

My chosen country is: _____

This country's form of money is:

_____

Fact 1:

_____

_____

_____

Fact 2:

_____

_____

_____

Books I used:
Title:

_____

Author:

_____

Title:

_____

Author:

_____

Title:

_____

Author:

_____

Money Glossary

Materials:

"Money Glossary" Hands-on Handout (p. 42), dictionaries, pencils, construction paper, stapler, crayons or markers

Directions:

1. Duplicate one "Money Glossary" Hands-on Handout for each child. Explain that a glossary is a list of special words with definitions listed after each word.
2. Have the children look up each word in the dictionary.
3. The children should write the definition next to the word to create their own money glossaries. They can put the definitions into their own words. This will help them to remember the definitions later. Younger children can draw pictures to represent the meanings of the words.
4. As children learn new money words or phrases, they can add these to their glossaries.
5. Provide construction paper and a stapler for the children to use to bind their pages together. They can decorate the covers of the books with drawings of different money icons.

Option:

• White-out the words on the "Money Glossary" handouts and duplicate one page for each child. Let the children write in their own money-related words and definitions.

Money Glossary

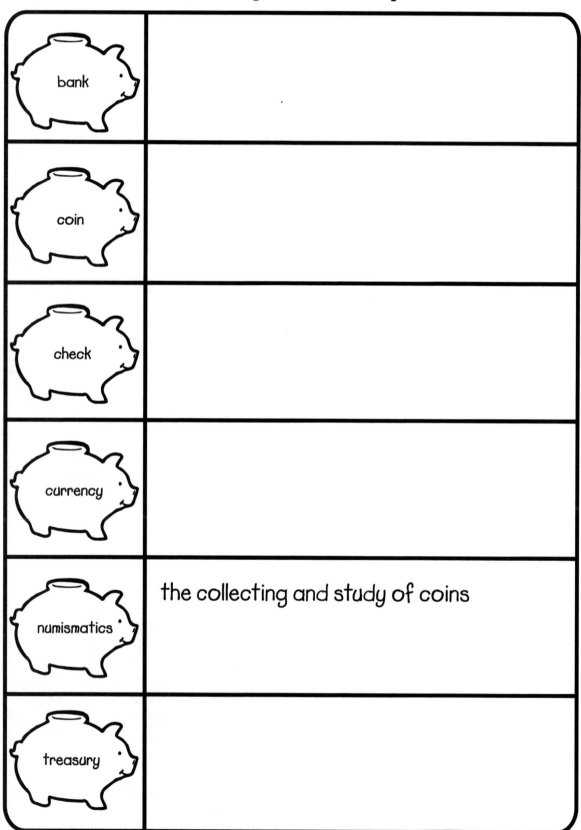

bank	
coin	
check	
currency	
numismatics	the collecting and study of coins
treasury	

Savings Spelling Bee

Materials:
Spelling Word Coin Patterns (pp. 44-45), Piggy Bank (p. 46), crayons or markers, hat, scissors

Directions:
1. Duplicate the "Spelling Word Coin Patterns," making one sheet for each child and a few extra sheets for teacher use.
2. Color the "Piggy Bank" and post it on a bulletin board. Cut out one extra set of coins and post them around the bank. (Cover the board during the spelling bee.)
3. Have the children learn how to spell each word. (You might give the children the sheets to take home and study at the beginning of the week, then have the spelling bee at the end of the week.)
4. Host a spelling bee in your classroom. Keep one set of coins in a hat and pull out one at a time, asking each chid in turn to spell the word on the coin. (Older children may be able to both spell and define the word.)
5. Continue with the spelling bee, having each child who misspells a word sit down.

Note:
Choose words that are appropriate to your children's spelling level. If a word is extra-hard for the students, it could be classified as a bonus term. A child who misspells it won't have to sit down, but would get another turn with a different word.

Options:
• Duplicate blank coins and let children write in their own money-related terms.
• Younger children can simply tape the coins to sheets of writing paper and practice copying the words.

Spelling Word Coin Patterns

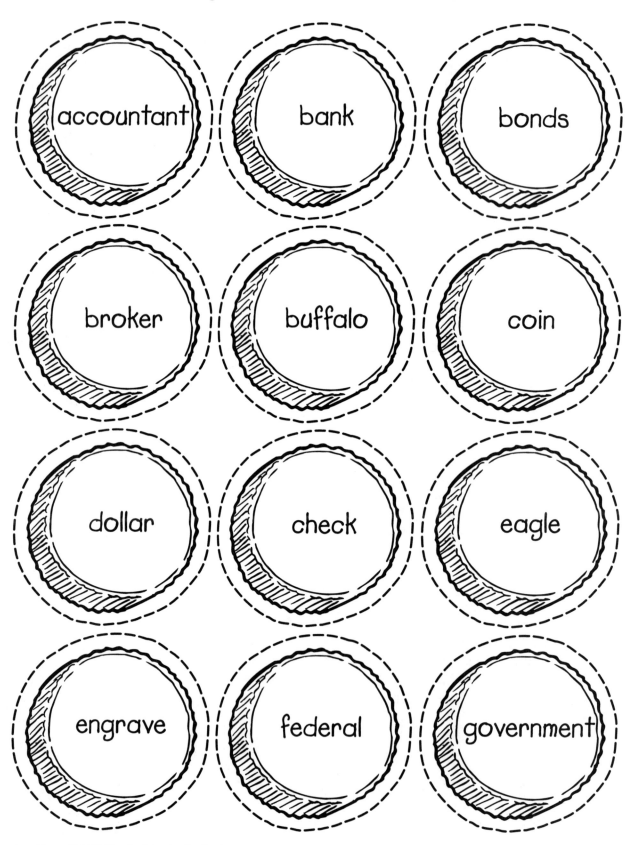

accountant

bank

bonds

broker

buffalo

coin

dollar

check

eagle

engrave

federal

government

Spelling Word Coin Patterns

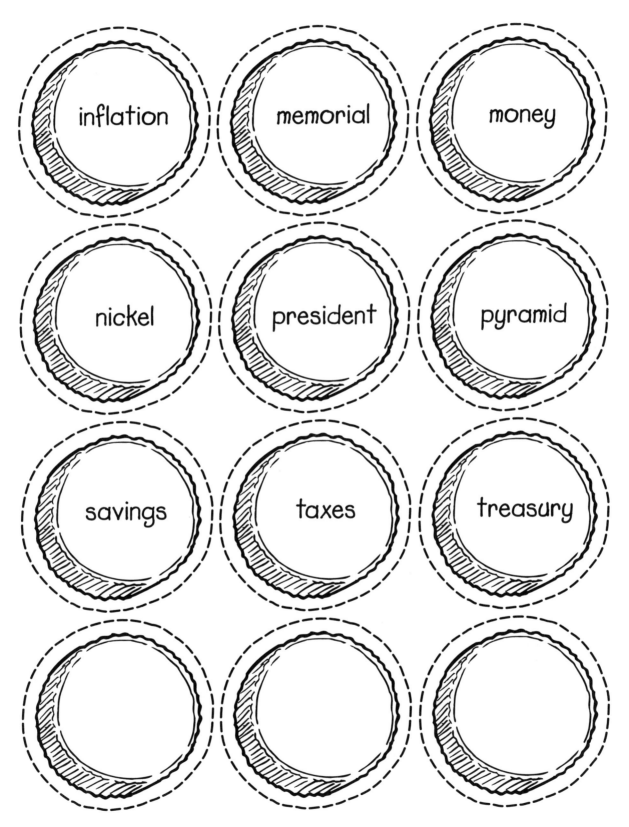

Marvelous Money © 2000 Monday Morning Books, Inc.

Piggy Bank

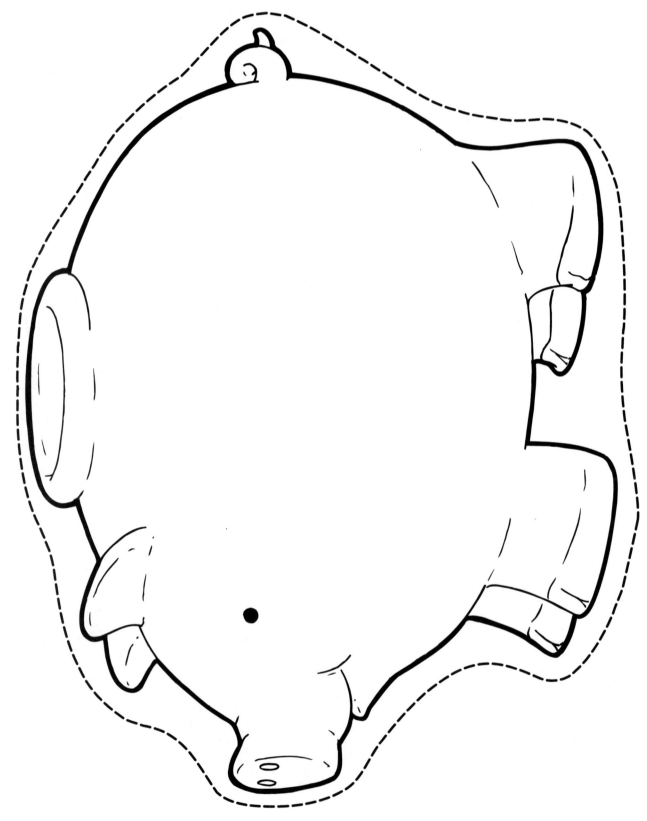

I'm on the Money

The children have studied the people on the front of money and the buildings on the back. Now, they can create a form of money that features their own pictures on the front, and their homes on the back.

Materials:
Photographs of the children's faces (brought from home), construction paper, scissors, notebook paper, crayons or markers, pens or pencils, glue, laminating machine or clear contact paper

Directions:
1. Discuss the different people who are on the money in your country.
2. Give the children paper, scissors, and markers to use to create paper currency.
3. Have the children glue their photos to the fronts of the bills and draw their homes on the backs.
4. Laminate the finished money to protect it.
5. On lined paper, have each child write a short autobiography explaining why he or she has been featured on a bill. For instance, a child might claim to be the first person on Mars, or the inventor of an automatic dog walking machine.
6. Let the children share their bills and their autobiographies with the class. Then post the bills and the autobiographies on a "We're on the Money" bulletin board.

Note:
Be sure to tell the parents that these pictures will be cut up and used in an assignment. They shouldn't send their only copies.

Marvelous Money © 2000 Monday Morning Books, Inc.

Charlie and the Chocolate Factory

Story:
Charlie and the Chocolate Factory by Roald Dahl, illustrated by Joseph Schindelman (Bantam, 1964). Charlie Bucket and his family hardly have enough money to eat. However, once a year, Charlie receives a ten-cent chocolate bar on his birthday. This year, the big chocolate factory in town has hidden golden tickets in five chocolate bars. The ticket-holders will win a visit into the factory. Charlie's birthday chocolate bar doesn't hold a winning ticket. But when Charlie finds a dollar on the street, he purchases a chocolate bar and finds lucky ticket number five. This begins his adventure in Willy Wonka's chocolate factory.

Setting the Stage:
• Put a golden (or yellow) ticket on each child's desk before the students arrive at school. The children can use the tickets as bookmarks during this unit.
• Create a candy factory feel in the classroom by cutting lollipops and hard candy shapes from colored construction paper. Post them on a bulletin board with chocolate bar shapes or real chocolate bar wrappers.
• Make desk labels, hall passes, or name tags on chocolate bar-shaped pieces of paper.
• Serve miniature chocolate bars as a treat.

Tricky Tongue Twister:
Chew sugar-dipped chocolate chips.

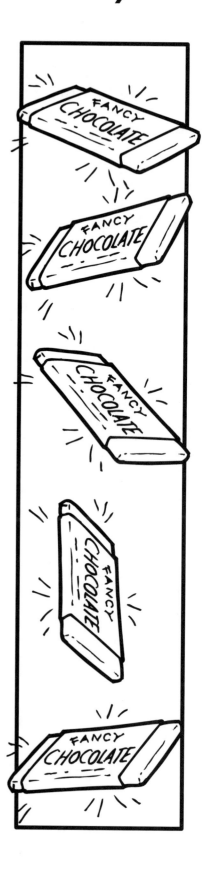

Marvelous Money © 2000 Monday Morning Books, Inc.

Creating a Golden Ticket

Charlie's golden ticket opens a whole new world for him. In this activity, the children will create golden tickets to their own fantasy locations.

Materials:
Yellow or gold construction paper, notebook paper, scissors, pens or pencils, glue, crayons or markers, glitter mixed with glue (stored in squeeze bottles)

Directions:
1. Have the children imagine that they have each opened a chocolate bar to discover a golden ticket. Now have them imagine that instead of a trip to a chocolate factory, their tickets will take them anywhere they want to go. They could hope for a trip to a real location like Paris, an imaginary place like Oz, a different time like the age of the dinosaurs, or a real place that no people have visited yet like Mars.
2. Give each child a sheet of yellow construction paper to turn into a golden ticket. They can decorate their tickets using crayons or markers and glitter mixed with glue.
3. While the tickets dry, have the children write a short paragraph about where their tickets will take them. They can do this on the notebook paper.
4. Post the completed tickets and written descriptions on a "Going for the Gold" bulletin board.

Option:
• Decorate the bulletin board with a variety of different colored tickets. These are often available at stationery stores, or have the children save them from any event their family attends: movies, concerts, the circus, and so on.

Marvelous Money © 2000 Monday Morning Books, Inc.

Rumpelstiltskin

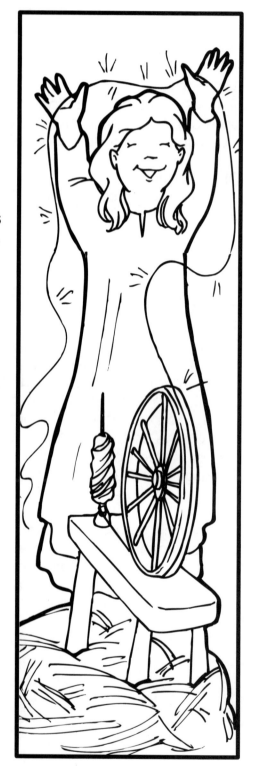

Story:
Rumpelstiltskin retold and illustrated by Paul Galdone (Clarion, 1985).

A miller brags to the king that his lovely daughter can spin straw into gold. The king puts the daughter to a test—if she cannot spin straw into gold she will die. A little man appears to help the daughter. The price of his help is her first-born child. When the time comes to collect the child, the girl doesn't want to give up her baby. The little man says that he will forgo the debt if she can guess his name. With help from a faithful servant, she does. Another version of this famous tale is retold and illustrated by Paul O. Zelinsky (Dutton, 1986).

Setting the Stage:
• Set real straw on a table (or straw created from shredded yellow paper or yellow pipe cleaners). Nearby, place large spools of gold or yellow-colored yarn. Post a picture of a spinning wheel on a bulletin board above the straw and "gold thread."

• Place name resources, such as baby name books or character naming guides, on a table for children to look through.

Tricky Tongue Twister:
Spin slick straw to shining threads.

Rumpelstilt-cents

Different countries feature different people or characters on their money. An Australian bill features the writer of the song "Waltzing Mathilda." In France, the ten-franc bill features a picture of the famous storybook character the Little Prince by Antoine de Saint Exupéry.

Materials:
Pictures of foreign money, paper, crayons or markers

Directions:
1. Show the children pictures of money from different countries. (See **Nonfiction Resources**.)
2. Have the children pretend that they are in charge of making money. If they could put any fictional character on a bill, who would they choose? Explain that there are no limits or right or wrong answers. They might choose a famous storybook character, such as the Cat in the Hat, or characters from fairy tales or fables, such as Rumpelstiltskin, Sleeping Beauty, or Hansel and Gretel. They might choose characters from video games or their favorite television shows.
3. Provide art materials for the children to use to create bills or coins featuring their chosen character.

Option:
• Have the children write about why their character should be on the next dollar bill. The children can then vote on which characters they would most like to have on their bills.

Penny in the Road

Story:
Penny in the Road by Katharine Wilson Precek, illustrated by Patricia Cullen-Clark (Macmillan, 1989).
This story takes place in 1913, when a little boy walking down a country road finds a penny in a puddle. The penny is from 1793, and the little boy imagines who might have dropped it. When a coin collector offers to trade the boy his special penny for a pocket knife, the boy refuses. Instead, he keeps it safe and shares it with his children and grandchildren throughout the years.

Setting the Stage:
• Set a variety of period-piece books on a table for the children to look through, such as *Yankee Doodle* illustrated by Steven Kellogg, *The Great Brain* series by John D. Fitzgerald, and *The Little House on the Prairie* series by Laura Ingalls Wilder.
• Decorate the bulletin board with pictures of pennies cut from brown or copper-colored paper.
• Host a costume day during which the children dress up in costumes from different time periods.

Tricky Tongue Twister:
Plenty of pennies, please.

Penny Wishes

The little boy in the story considers his 1793 penny to be worth even more than the pocket knife he's been wanting. Maybe he thinks this penny is lucky.

Materials:
"Lucky Coin" Hands-on Handout (p. 54), pens or pencils, scissors

Directions:
1. Duplicate a copy of the "Lucky Coin" Hands-on Handout for each child.
2. Have the children follow the directions on the page to write short stories about good luck.
3. When the children are finished, create a wish pond on a bulletin board using construction paper and crepe paper.
4. Have the children cut out the coin shapes that their stories are written on. Then post the children's lucky coin stories on the bulletin board.
5. When you take down the display, you can bind the pennies in a good luck book.

Options:
• Have children write lucky stories about any type of coins: pennies, nickels, dimes, or quarters. Perhaps a quarter would bring 25 days of good luck and a nickel would bring five days of good luck.
• Teach the children the following rhyme:
> Find a penny, pick it up,
> And all day you'll have good luck!

Lucky Coin

What You Do:
Imagine that you have found a lucky penny. What type of luck might it bring for you? Write a very short story about finding a lucky penny.

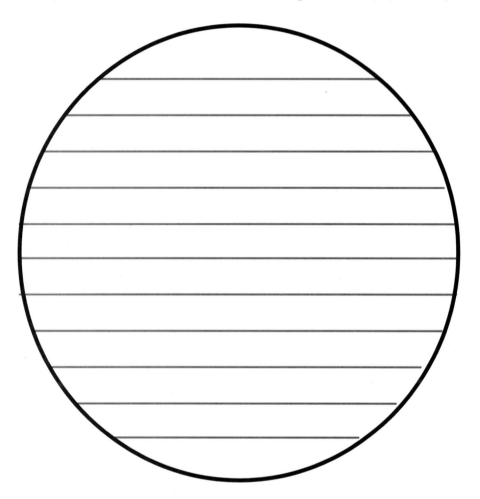

Just for Fun: Lucky Charms
People all over the world consider finding coins to be good luck. Start a list of all "good luck charms" that you can think of. Ask your friends if they have any objects that they consider lucky. You might be surprised!

The Money Tree

Story:

The Money Tree by Sarah Stewart, illustrated by David Small (Farrar, Straus & Giroux, 1991).
A woman wakes up to discover a small green tree growing in her yard. As the seasons change, the tree continues to grow. However, it's not like any tree she's seen before. It has a very strange shape and it grows more quickly than the rest of her plants. The strangest thing about the tree is its leaves. Instead of being normal green leaves, this tree has money for leaves! Eventually, the woman has to come to a decision of what to do with the strange money tree.

Setting the Stage:

• Decorate a bulletin board with fallen leaves or leaves cut from different colors of construction paper.
• Create a money tree by posting a brown paper tree trunk on a bulletin board and pasting on green bill-shapes for the leaves.
• This book covers an entire year. Focus on the part of the book that features the current season. For instance, if you read the book in October, post pictures of jack-o'-lanterns around the classroom. Or, in May, let the children create maypoles from ribbons, pipe cleaners, and clay bases.

Tricky Tongue Twister:

How much money does a mound amount to?

Money Tree Clichés

In order to be good writers, it's important for your students to avoid clichés. They should be able to describe things in ways that other people haven't.

Materials:
"Money Doesn't Grow on Trees" Hands-on Handout, pencils or pens

Directions:
1. Duplicate a copy of the "Money Doesn't Grow on Trees" handout for each child.
2. Discuss the concept of clichés with the children.
3. Have the children follow the directions on the handout.
4. Let the children share their rewritten clichés with each other.

Option:
• Whenever your students recognize a cliché, have them write it down in a notebook. They can collect a list of clichés to avoid. Next to each cliché, have them write a new way to say the same thing. For example, instead of saying "it's raining cats and dogs," they might write "it's raining cows and buffalos."

Money Doesn't Grow on Trees

What You Do:

1. Has anyone ever told you that money doesn't grow on trees? Often people get used to saying the same thing over and over. Then the statement can become a cliché. Below are several different examples of clichés:

Money doesn't grow on trees.
Never look a gift horse in the mouth.
A stitch in time saves nine.
A penny saved is a penny earned.
It's raining cats and dogs.

2. Think about the above statements. Do they make sense? Rewrite one of the clichés above (or any other cliché that you know). Make the new statement more modern.

Marvelous Money © 2000 Monday Morning Books, Inc.

Pigs Will Be Pigs

Story:
Pigs Will Be Pigs by Amy Axelrod, pictures by Sharon McGinley-Nally (Four Winds Press, 1994).
The pig family is hungry, but they don't have any food in the refrigerator and they are out of money. They decide to hunt for the money, looking for change all over the house. When they collect enough, they go to a restaurant for dinner. This book poses several math problems. It tells the children how much money the pigs have and then features a menu for the children to use to figure out which dishes the pigs can afford. Additional math problems are posed at the end of the book.

Setting the Stage:
• Bring in menus from a variety of restaurants. Post the menus around the room. The children can use play money to pretend to buy various meals.
• Post pictures of pigs around the room.
• Serve a Mexican food snack, something the pigs might have ordered at the Enchanted Enchilada.

Tricky Tongue Twister:
Pretty pigs are purely pink.

The Perfect Menu

Materials:
Menus, crayons or markers, paper

Directions:
1. Bring in menus from local restaurants for the children to observe.
2. Explain that the children will be creating menus for imaginary restaurants. Have them each brainstorm a list of their favorite foods.
3. Provide crayons and markers and paper for the children to use to make menus. The menus for their imaginary restaurants should feature all of their favorite things to eat. Have the children write the names of their restaurants on the tops of the menus.
4. Have the children list a price for each meal on the menu.
5. Let the children exchange menus with each other. The partners can pretend to have enough money for three items on the menu. They can then tell each other which items they would choose.
6. Post the completed menus in the classroom.

Option:
• Consider letting the children each bring in a favorite food item to share with the rest of the students.

Book Link:
• *Yoko* by Rosemary Wells (Hyperion, 1998).

Alexander, Who Used to Be Rich Last Sunday

Story:
Alexander, Who Used to Be Rich Last Sunday by Judith Viorst, illustrated by Ray Cruz (Atheneum, 1978). Alexander and his two brothers all like money. Still, Alexander has nothing but bus tokens at the start of this book, but he remembers a week ago when he had a whole dollar. Although his mother told him he should save his dollar toward the walkie-talkie he wants, throughout the week, he spends his pennies, nickels, and dimes on various items.

Setting the Stage:
• Make a bulletin board featuring the different items that tempt Alexander throughout the story, such as gum, a candle, a used teddy bear, and an hour with his friend's snake.
• Place the other books about Alexander on a table where the children can look through them. These are *Alexander and the Terrible, Horrible, No Good, Very Bad Day* and *Alexander Who's Not (Do You Hear Me? I Mean It!) Going to Move.*
• Use a calendar to chart the way Alexander spent his money throughout the book, from one Sunday to another.

Tricky Tongue Twister:
Surely she saves. She saves some.

If I Had a Dollar

Alexander had a dollar, but he spent it. In this activity, the children will write a story about how they might have spent the dollar differently.

Materials:
Writing paper, pencils, drawing paper, crayons or markers

Directions:
1. Have the children imagine that they are in Alexander's place and they have each been given a dollar.
2. Provide writing paper and have the children write short stories about how they might have spent their dollar in ways different from Alexander. (If they agree with the way he spent it, they can write that.)
3. Let the children illustrate their stories using drawing paper and crayons or markers.
4. Bind the stories in an "If We Had a Dollar" storybook, or post the stories and pictures in the classroom.

Options:
• If you post the stories, make a border using the Money Patterns (p. 30).
• Because *Alexander, Who Used to Be Rich Last Sunday* was written in 1978, consider upping the dollar amount to five or ten dollars for this activity. Decide on an amount that your students can relate to.

Marvelous Money © 2000 Monday Morning Books, Inc.

Piggy Bank Stories

Materials:
Piggy Bank (p. 46), lined paper, pencils or pens, crayons or markers, hole punch, yarn, tagboard or oak tag

Directions:
1. Three different stories discussed in this chapter deal with spending money in one form or another. Charlie buys the golden ticket, the pigs go out to dinner, and Alexander buys a wide variety of items. Have the students imagine that they have some money—the amount doesn't matter—but that instead of spending it, they are going to save it for something special.
2. Duplicate the Piggy Bank pattern onto lined paper. Make one piece for each child in the class. Duplicate two extra copies onto thick tagboard or oak tag.
3. Have the children write short stories about the money they're saving in their piggy banks and what they are saving for.
4. Bind the stories in a classroom "Piggy Bank" Album. Make a cover for the book using the patterns duplicated onto heavy paper. Punch two holes in each page and bind together with yarn.
5. Display the book in your classroom.

Marvelous Money Program

Songs:
- Jingling Coins
- A Penny's Worth One Cent
- Abe Lincoln Once Led Our Country
- Jefferson Is on the Nickel
- Check Your Nickels, Now
- Roosevelt Is on the Dime
- A Dollar's Worth a Lot
- We Learned How to Spell "Quarter"

Featuring:

Jingling Coins

(to the tune of "Jingle Bells")

Jingling coins, jingling coins,
    in my pocketbook.
How much money do I have?
I'll have to take a look.
Quarters, dimes, nickels, too,
They make a pretty sound.
I'll add them all together
    with this penny that I found.

Dimes are ten cents each.
Nickels equal five.
Pennies are worth one,
And quarters twenty-five.

I'll add them one by one,
And put them safe away.
Saving money can be fun,
For a cold or rainy day!

I'll add them up one by one,
And put them safe away.

A Penny's Worth One Cent

(to the tune of "My Country 'Tis of Thee")

A penny's worth one cent.
It has a president,
Right on its face.
His name was Abraham,
He was an honest man,
So save your pennies when you can,
In your safest place.

Abe Lincoln Once Led Our Country

(to the tune of "My Bonnie Lies Over the Ocean")

Abe Lincoln once led our country,
He was our sixteenth president.
And now we have honored his memory,
By putting him on the one-cent, one-cent.
Lincoln's on pennies and also on five-dollar bills,
    yes, bills.
Lincoln's on pennies and also on five-dollar bills.

Jefferson Is on the Nickel

(to the tune of "My Bonnie Lies Over the Ocean")

Oh, Jefferson is on the nickel,
And Roosevelt is on the dime,
Abe Lincoln is featured on pennies,
So save up your change every time, each time.
Save all your pennies, your nickels, and quarters
    and dimes, and dimes.
Save all your pennies, your nickels,
    and quarters and dimes.

Check Your Nickels, Now

(to the tune of "Row, Row, Row Your Boat")

Check your nickels now,
Do you know that man?
Jefferson, Jefferson, Jefferson, Jefferson,
He once led our land.

Marvelous Money © 2000 Monday Morning Books, Inc.

Roosevelt Is on the Dime

(to the tune of "Mary Had a Little Lamb")

Roosevelt is on the dime,
On the dime,
On the dime,
Roosevelt is on the dime.
It's worth ten cents.

A Dollar's Worth a Lot

(to the tune of "My Country 'Tis of Thee")

A dollar's worth a lot,
More than you might have thought,
One hundred cents.
It's also worth ten dimes,
Four quarters in a line,
Or twenty nickels, see them shine?
Saving makes good sense.

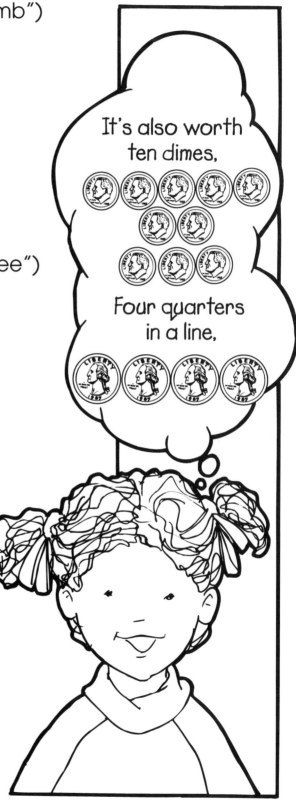

Marvelous Money © 2000 Monday Morning Books, Inc.

We Learned How to Spell "Quarter"

(to the tune of "Take Me Out to the Ball Game")

We learned how to spell quarter,
Quarter starts with a q—
So do words like quail and quarterback,
Quake, quit, quartz, queen, quill,
and quack-quack-quack!

More "q" words are quilt and quiet,
And question mark, quicksand and quote,
There are lots of "q" words out there,
So you all take note.

Money Music

Songs have been written about many different kinds of things. Your students can write songs about money to help them remember facts they've learned. Or they can start by writing songs about any themes.

Materials:
Paper, pens or pencils, songs from this unit

Directions:
1. Duplicate one or more of the songs from this unit. Make one copy for each student.
2. On the chalkboard, write the original words to the chosen song or songs.
3. Explain to the children that these songs were written by putting new words to familiar tunes.
4. Have the children write new songs to familiar tunes. They should choose very simple songs to start with, such as "Row, Row, Row, Your Boat," "Jingle Bells," or "Clementine." Remind the children that if they add facts to their songs, it may help them to remember the facts later. Be sure to keep encouraging the students. If their songs don't work out at first, they should keep trying, or choose a new song to write words to.
5. When they children are done, have them name their songs and share them with the class.

Options:
• Have the children work together in song-writing teams.
• Tape the children's songs to share with parents or other classes.

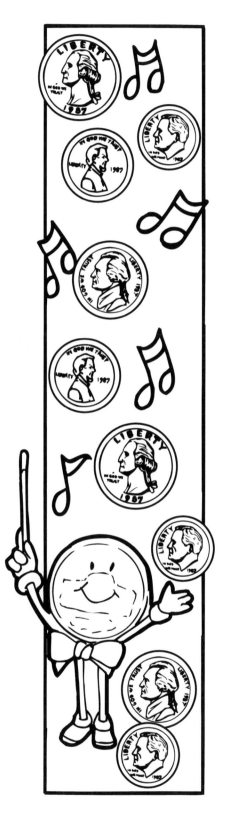

Marvelous Money © 2000 Monday Morning Books, Inc.

Abraham Lincoln Facts

Money: Abraham Lincoln is on the penny and the five-dollar bill.
President: He was the 16th president of the United States.
Lifetime: Lincoln lived from 1809 to 1865.

Building: The Lincoln Memorial is on the back of both the penny and the five-dollar bill. It was built between 1911 and 1922. It is in West Potomac Park, Washington, D.C.
Super-Duper Fact: Lincoln is famous for signing the Emancipation Proclamation. This document freed the slaves.

George Washington Facts

Money: George Washington is on the dollar bill and the quarter.
President: He was the first president of the United States. George Washington fought for American freedom.

The Great Seal: On the back of the one-dollar bill is the Great Seal of the U.S. It features an eagle holding arrows and an olive branch. The back also shows a pyramid.
Lifetime: Washington lived from 1732 to 1799.
Super-Duper Fact: George Washington did not go to college.

Thomas Jefferson Facts

Money: Thomas Jefferson is on the nickel. He is also on two-dollar bills.
President: Jefferson was the third president.
Other Skills: He was an inventor and an architect.
Building: On the back of the nickel is a building called Monticello. It was Jefferson's home in Charlottesville, Virginia. Monticello means "little mountain" in Italian.
Lifetime: Jefferson lived from 1743 to 1826.
Super-Duper Fact: Jefferson wrote the Declaration of Independence in 17 days!

Franklin D. Roosevelt Facts

Money: Franklin D. Roosevelt is on the dime.
President: Roosevelt was the 32nd president.
Hobby: Roosevelt collected stamps.
Building: On the back of the dime is the liberty torch and oak and olive branches.
Lifetime: Roosevelt lived from 1882 to 1945.
Super-Duper Fact: Roosevelt's favorite food was frogs' legs.

John F. Kennedy Facts

Money: John F. Kennedy is on the 50-cent piece.
President: Kennedy was the 35th president.
Education: Kennedy graduated from Harvard University.

Prizes: Kennedy wrote a book called *Profiles in Courage*. It won the Pulitzer Prize for biography.
Lifetime: Kennedy lived from 1917 to 1963.
Super-Duper Fact: Kennedy's favorite books were in the *James Bond* series.

Alexander Hamilton Facts

Money: Alexander Hamilton is on the ten-dollar bill.
Occupation: Hamilton was the first secretary of the treasury.
Building: On the back of the ten-dollar bill is the U.S. Treasury

Building. There have been several treasury buildings. The first two were destroyed by fire.
Lifetime: Hamilton lived from 1755 to 1804.
Super-Duper Fact: The current treasury building is the oldest government department building in Washington. It is next to the White House on Pennsylvania Avenue.

Andrew Jackson Facts

Money: Andrew Jackson is on the 20-dollar bill.
President: Jackson was the seventh president.
Occupation: Jackson was also a soldier. He became a general in the War of 1812.
Building: On the back of the 20-dollar bill is the White House. It was originally built in 1792, rebuilt in 1817, and restored between 1948 and 1950.
Lifetime: Jackson lived from 1767 to 1845.
Super-Duper Fact: Jackson was a war hero.

Ulysses S. Grant Facts

Money: Ulysses S. Grant is on the 50-dollar bill.
President: Grant was the 18th president.
Building: On the back of the 50-dollar bill is the Capitol.
The United States Capitol is located on Capitol Hill.
Food: Grant's favorite food was cucumbers soaked in vinegar.
Lifetime: Grant lived from 1822 to 1885.
Super-Duper Fact: Grant was the first soldier to reach the rank of "General in Chief of the Union Forces" since George Washington.

Benjamin Franklin Facts

Money: Benjamin Franklin is on the 100-dollar bill.
Occupation: Franklin was a printer, inventor, statesman, and postmaster.
Building: Independence Hall is on the back of the 100-dollar bill.

Lifetime: Franklin lived from 1706 to 1790.
Super-Duper Fact: Franklin was the only American to have signed these four key American documents: the Declaration of Independence, the Treaty of Alliance with France, the Treaty of Paris, and the Constitution.

Susan B. Anthony Facts

Money: Susan B. Anthony is on a dollar coin.
Occupation: Anthony was a feminist leader.
Coin: The coin was issued in 1979.
Lifetime: Anthony lived from 1820 to 1906.

Good Deeds: Anthony helped secure the first laws in New York that guaranteed women's rights over their children and control of property and wages.
Super-Duper Fact: People didn't like using the coin because it felt like a quarter. It is no longer made. More than 400 million Susan B. Anthony dollars are in storage.

William McKinley Facts

Money: William McKinley is on the 500-dollar bill.
President: McKinley was the 25th president.
Occupation: McKinley was also a soldier and a statesman.

He was second lieutenant, 23rd Ohio Volunteer Infantry during the Civil War.
Lifetime: McKinley lived from 1843 to 1901.
Super-Duper Fact: McKinley was assassinated in September of 1901.

Grover Cleveland Facts

Money: Grover Cleveland is on the 1000-dollar bill.
President: Cleveland was the 22nd and 24th president. He was the only president to serve non-consecutive terms.

Good Deeds: Cleveland signed the Yellowstone Act. This was the first Federal legislation to protect wildlife on government lands.
Lifetime: Cleveland lived from 1837 to 1908.
Super-Duper Fact: Cleveland was the first president to be married in the White House.

James Madison Facts

Money: James Madison is on the 5000-dollar bill.
President: Madison was the fourth president.
Lifetime: Madison lived from 1751 to 1836.

The First Lady: Dolly Madison, James Madison's wife, saved state papers from the White House when it was burned by the British in the War of 1812.
Super-Duper Fact: Madison is known as "The father of the United States Constitution."

Salmon P. Chase Facts

Money: Salmon P. Chase is on the 10,000-dollar bill.
Occupation: Salmon P. Chase was the 25th secretary of the treasury.
Lifetime: Chase lived from 1808 to 1873.
President: Chase was never president, but he wanted to be. He tried several times to be nominated by his party, but he always lost.
Super-Duper Fact: Salmon P. Chase ordered "In God We Trust" to appear on all coins minted during the Civil War. Today, this statement is on all U.S. currency.

Woodrow Wilson Facts

Money: Wilson is on the 100,000-dollar bill.
President: Wilson was the 28th president.
Occupation: Wilson was also an educator.

Education: Wilson attended graduate school.
Foods: Wilson's favorite foods were grapefruit juice and raw eggs.
Lifetime: Wilson lived from 1856 to 1924.
Super-Duper Fact: Wilson won the Nobel Peace Prize.

Treasurer Facts

Getting the Job: The President of the U. S. appoints the treasurer.
Duties: The job includes overseeing the Bureau of Engraving and Printing, the U.S. Mint, and the United States Savings Bond Division. The treasurer signs his or her name on the lower left side of all U.S. currency.
Counterfeiters: The treasurer also is the chairperson for the Advanced Counterfeit Deterrence Steering Committee.
Super-Duper Fact: There have been male and female treasurers.

Marvelous Money © 2000 Monday Morning Books, Inc.

Money A to Z List

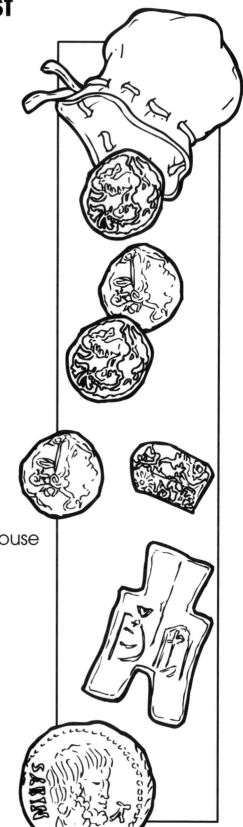

A: Anthony, Susan B.
B: Bank, Barter
C: Coin, Currency
D: Dime, Dollar
E: Eagle
F: Feathers
G: Gold
H: Half-dollars
I: Inflation, Ingot
J: Jefferson, Thomas
K: Krona (Swedish monetary unit)
L: Lincoln, Abraham; Lira
M: Monticello
N: Nickel
O: Owl coins (the silver coins of Athens)
P: Penny
Q: Quarter
R: Rupee (Indian monetary unit)
S: Salt, Silver
T: Trade
U: United States Treasury
V:
W: Wampum; Washington, George; the White House
X:
Y: Yen (Japanese monetary unit)
Z: Zloty (Polish monetary unit)

Nonfiction Resources

Money Books:
• *All Kinds of Money* by David A. Adler, illustrated by Tom Huffman (Franklin Watts, 1984).
• *Banks: Where the Money Is* by David A. Adler, illustrated by Tom Huffman (Franklin Watts, 1985).
• *The Go-Around Dollar* by Barbara Johnston Adams, illustrated by Joyce Audy Zarins (Four Winds Press, 1992).
• *Inflation: When Prices Go Up, Up, Up* by David A. Adler, illustrated by Tom Huffman (Franklin Watts, 1985).
• *Making Cents: Every Kid's Guide to Money* by Elizabeth Wilkinson, illustrated by Martha Weston (The Yolla Bolly Press, 1989).
• *Money* by Joe Cribb (Knopf, 1990).
• *Money, Money, Money* by Nancy Winslow Parker (HarperCollins, 1995).
• *Neale S. Godfrey's Ultimate Kids' Money Book* by Neale S. Godfrey, illustrated by Randy Verougstraete (Simon & Schuster, 1998).
• *The Story of Money* by Betsy Maestro, illustrated by Giulio Maestro (Clarion, 1993).

Pyramid Books:
• *Ancient Egypt* by Rosalie and Antony E. David (Warwick, 1984).
• *Exploring Ancient Egypt* by John Malam (Evans Brothers, 1997).
• *Pyramid* by James Putnam (Knopf, 1994).

Nonfiction Resources

Web Sites:
• http://www.fortress.am/users/dollar/#sba
This site shows a picture of the Susan B. Anthony coin, as well as the Eisenhower silver dollar and several other coins.
• http://www.susanbanthonyhouse.org/dollar.htm
This site features information about the Susan B. Anthony coin.
• http://www.usmint.gov/dollarcoin/
This site shows a picture of the Susan B. Anthony coin and also features the new coin commemorating Sacagawea, who assisted in the Lewis and Clark expedition.
• http://www.usmint.gov/default.htm
This is the main Web site of the U.S. Mint.

Note:
The best efforts have been made to find current Web sites. However, Web sites sometimes change. In addition to using these sites, also try keyword searches, such as specific types of money.

Game That Uses Money:
• Monopoly

Play Money:
Packages of play U.S. coins and U.S. currency are available at teacher supply stores, as well as some office supply stores.

Money Stickers:
Stickers featuring U.S. money (bills and coins) are available from Mrs. Grossman's.

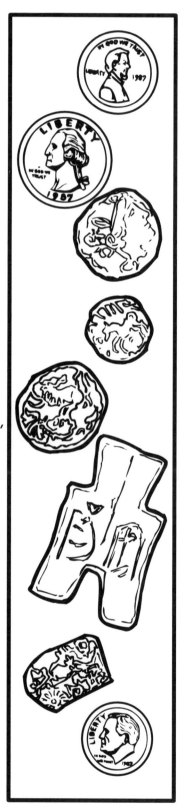